What a Truly Cool World

by JULIUS LESTER
illustrated by JOE CEPEDA

SCHOLASTIC INC.
New York Toronto London Auckland Sydney
Mexico City New Delhi Hong Kong

To Claudia Menza,
my friend —JL

For David and
Denna —JC

In 1969, I published *Black Folktales*, my first book retelling African-American and African folktales. In that book, I refashioned a story called "How God Made the Butterflies," which I had encountered in Zora Neale Hurston's classic collection *Mules and Men*, a book long out of print at the time. My intention in *Black Folktales* was to reintroduce black folktales that were in danger of being forgotten. It was important that my retellings were faithful to the traditional tales and their story lines, and that my innovations were primarily of language and imagery. My purpose was to represent a people and their collective experience.

What A Truly Cool World is a departure for me from this kind of storytelling. Though "How God Made the Butterflies" was the starting point for this story, there is little resemblance between the two. Though the black storytelling voice is the same, here the characterization of God is fuller, and two new characters are introduced, the angels Bruce and Shaniqua. In other words, I allow my imagination free play here, something I did not do in the previous work because of my obligation to tradition. Now I want to be faithful to the infinite world of the imagination and the profound pleasures that world has to offer this one, which is perhaps the truest way of being faithful to tradition.

Julius Lester · Belchertown, Massachusetts · January 3, 1997

ISBN 0-590-86469-6

12 11 10 9 8 7 6 5 4 3 2 1 0 1 2 3 4 5/0

Printed in the U.S.A. 08

First Scholastic paperback printing, January 2000

The display type was hand lettered by Joe Cepeda.
The text type was set in Smile and Party.
The illustrations in this book were painted in oil.
Book design by Marijka Kostiw

When God finished making the world he felt as bright and sunny as love. He had never made a world before and, if he said so himself (and he did), he thought he had done a very good job.

God looked at the round ball of green and blue and brown suspended in the darkness of space. The green was the trees, the blue was water, and the brown was land. People were down there, too — black, white, yellow, and red ones, and animals — horned, clawed, winged, and furry. God smiled and decided to go home and get ready for bed.

Just then, Shaniqua, the angel in charge of everybody's business, came bursting through the door of the throne room.

"God? What you call that down there?" she wanted to know, pointing out the window.

"That's the earth! It's pretty, ain't it?"

Shaniqua looked at God like he was blind in one eye and deaf

in the other. "Don't look like much to me. I don't want to hurt your feelings or nothing like that, but what you made looks kind of boring."

God narrowed his eyes and stared at the world again. "You right, Shaniqua," he admitted reluctantly.

"I know I'm right! I'm always right!"

The next morning God woke up ready to do some more creating, but he didn't know what. When he walked into the throne room, the Hallelujah Angelic Choir of sixtillion voices started chanting like it had done every morning for the past ninetyleven tillion years: "God! God! He's our man! If he can't do it, nobody can!"

That's true, God thought. I am the man!

"Bruce!" he called out.

Bruce was God's secretary. He was standing in the corner brushing his wings.

"Yo! What's up, Deity?" he asked, hurrying over to God.

"Bring me a pair of scissors!"

Bruce had no idea what *scissors* were since they hadn't been invented yet. He flew off to the Library of Everything That Is Going to Be and typed "scissors" on the computer.

POOF!

Next thing he knew he was holding two long, sharp pieces of metal joined in the middle.

"Awesome!" Bruce proclaimed.

He flew back to the throne room. "Here you go, Mighty One!"

God leaned out the window, his arms stretching until they reached earth, and he began clipping the trees. Tiny green pieces of leaves floated down to the ground, and that was how grass and bushes came to be.

"That's better," God declared. "Now the world isn't so brown. That ought to satisfy Shaniqua."

God wasn't finished being pleased with himself before Shaniqua charged into the throne room.

"How's that?" God asked proudly.

Shaniqua was not impressed. "Blue, green, and brown. Is that the best you can come up with?"

God sighed and looked at the world again. Shaniqua was right. Everything still looked drab. What was he going to do?

The next morning God woke up with an idea, but he wasn't sure it would work. He skipped breakfast and headed toward the edge of heaven.

"What's up, Mighty Maker?" Bruce asked, flying alongside God.

"You'll see! Where's Shaniqua?"

"Right behind you, God."

The angels sensed something big was going to happen. By the time God arrived, they were sitting on the clouds, waiting. They had brought picnic baskets and coolers filled with lemonade and they were eager to see what God was going to do.

God stepped to the edge, cleared his throat, and out came the most beautiful sound the angels had ever heard.

"What's that?" Shaniqua wanted to know.

"Why, that's — that's music!" Bruce exclaimed. He had been reading about it in the Library of Everything That Is Going to Be. "That's MUSIC!" he repeated in wonder.

God's singing was like a choir of six million voices — high voices, low voices, voices singing notes only dogs heard, and voices singing notes only the whales in the deepest part of the ocean heard.

As the notes came out of God's mouth, they took shape and became circles and squares and spirals and trapezoids and rhombuses and triangles. The shaped notes floated through space and down to earth and when they touched the ground, they turned into colors — reds and blues and purples and yellows and pinks.

"What's all them colors?" an angel asked Bruce and pointed down at the earth.

Bruce shook his head slowly in amazement. "Those are flowers!"

"Flowers," the angels repeated in hushed whispers.

Shaniqua couldn't believe what she was hearing and seeing. "Go on with your bad self, God!" she laughed with delight.

As the notes sprinkled into the ground, they bloomed and each said its name and the names were like music:

Daff-o-dil

Sun-flower

Chic-o-ry

A-ma-ranth

Let me hear you say that:

A-ma-ranth. That's almost as pretty as God's singing, ain't it?

Cha-mo-mile

Gar-de-ni-a

Ba-by's Breeeaaath

Close to evening, God stopped and looked at Shaniqua.
"Give me five!" he said to her, holding out his hand.
She slapped it hard. "You are truly cool, God. Truly."

That night when God went to sleep, he slept soundly. Finally, he was done with creating the world.

The next morning God didn't have one eye open before he heard whispering.

"It's lonely down here."

It was the flowers.

"I know we were put here to make the world beautiful, but we're lonely."

God couldn't believe his ears. Maybe creating a world had been a big mistake. He was probably going to be hearing nothing but complaints from now until.

"Bruce!"

God called in a voice so loud that heaven shook.

Bruce had been trying to explain to his wife why he hadn't brought her any flowers yesterday when he heard God's voice.

"Oooops! Got to go. Himself is not in a good mood today. Don't wait dinner." He put on his wings and was out the window.

When Bruce reached God's house, the Almighty was sitting on the porch drinking coffee with his wife, Irene God. Bruce had always wondered what God's first name was, but he had never had the nerve to ask.

"Good morning, Terrific Titan."

"Ain't nothing good about it."

"What's the problem?"

"It's them flowers I made yesterday. They're lonely."

"Well, sing something into creation to keep them company," Bruce said and underneath his breath added, "George."

"My voice is hoarse from all that singing I did yesterday. And what did you just call me?" God asked, getting angry.

"Uh — nothing, Fearless Fixer. Nothing."

"This is not the morning to mess with me."

"Yes, sir."

"Where's Shaniqua?"

"She liked your singing so much that she's at home trying to teach herself to sing."

"How does she sound?"

"Well, to tell the truth, almost as good as you."

God smiled. "Go get her. I want to hear this for myself."

A few minutes later Bruce returned with Shaniqua. She had never been to God's house or met his wife, and she was as proper as a little girl playing grown-up.

"Bruce says you can sing almost as good as me," God said.

Shaniqua blushed and shook her head.

"No, sir. I just be singing around the house, that's all. Singing to myself, really."

"Well, let me hear you."

Shaniqua opened her mouth. The sound that came from her throat was as soft and as smooth as a flower's breath. As quiet as it was, it could be heard all over heaven. The angels stopped what they were doing to listen. The sound was so beautiful that water began to flow from their eyes. And that is how tears came into creation.

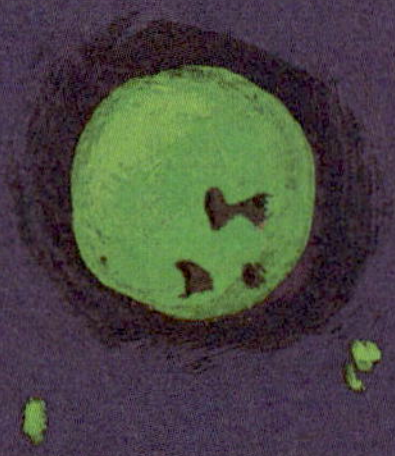

The music went out into the universe. The stars and the planets had never heard anything so beautiful. Tiny pieces started falling off of them. These were their tears.

The pieces floated down through space until they reached earth. People looked up at the tiny scraps fluttering around.

The yellow ones were the sun's tears.

The blue ones were tears from the moon.

The white ones were star tears,

while the red ones were from Mars

and the orange ones from Venus.

"What are they?" people wanted to know.

"Flutterbys," somebody called them.

The air was filled with flashing colors as the flutterbys flew

here and there

visiting and kissing each other.

"Look what you done done," God said to Shaniqua. "You created flutterbys."

"Wrong," said Bruce. "That silly word is never going to exist."

"Well, what are they?" God wanted to know.

Shaniqua smiled. "I know. The people are a little confused. What they mean to say is butterflies. Ain't that pretty, Irene?"

Mrs. God agreed that it was. So did God. So did the people on earth. They held out their arms and the butterflies covered them and their entire bodies until each person was wearing a suit of butterflies.

"Give me five," Shaniqua said to God.

He shook his head. "I couldn't have made the world without you, Shaniqua." And God gave her a big hug.

"What a truly cool world, God! Truly."